# Sea Jellies

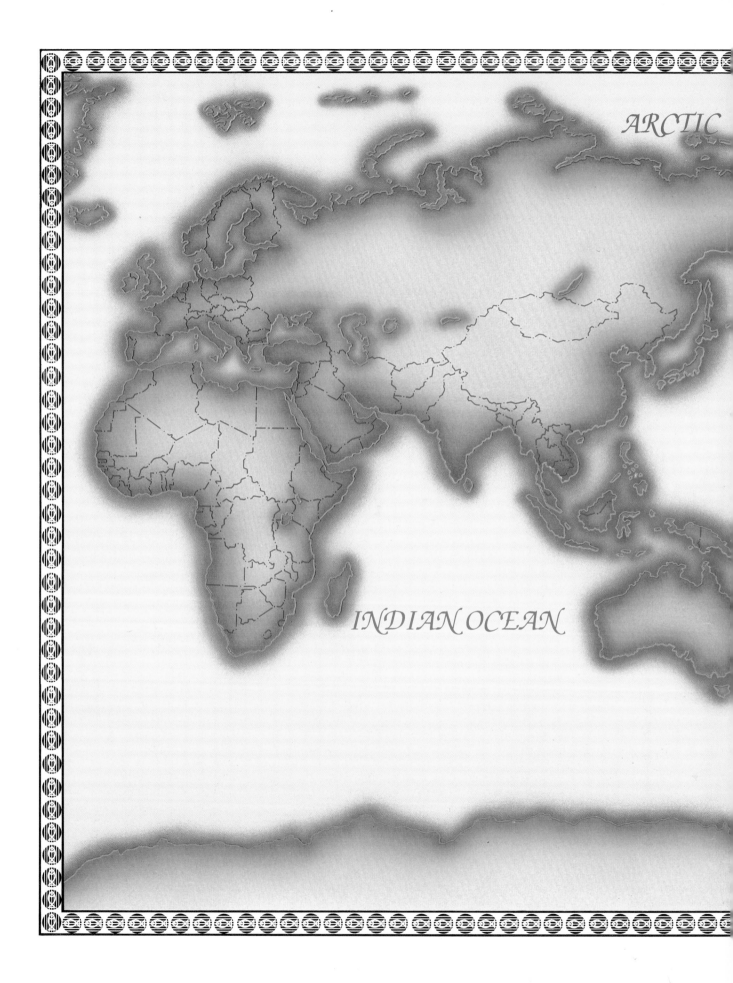

ARCTIC

INDIAN OCEAN

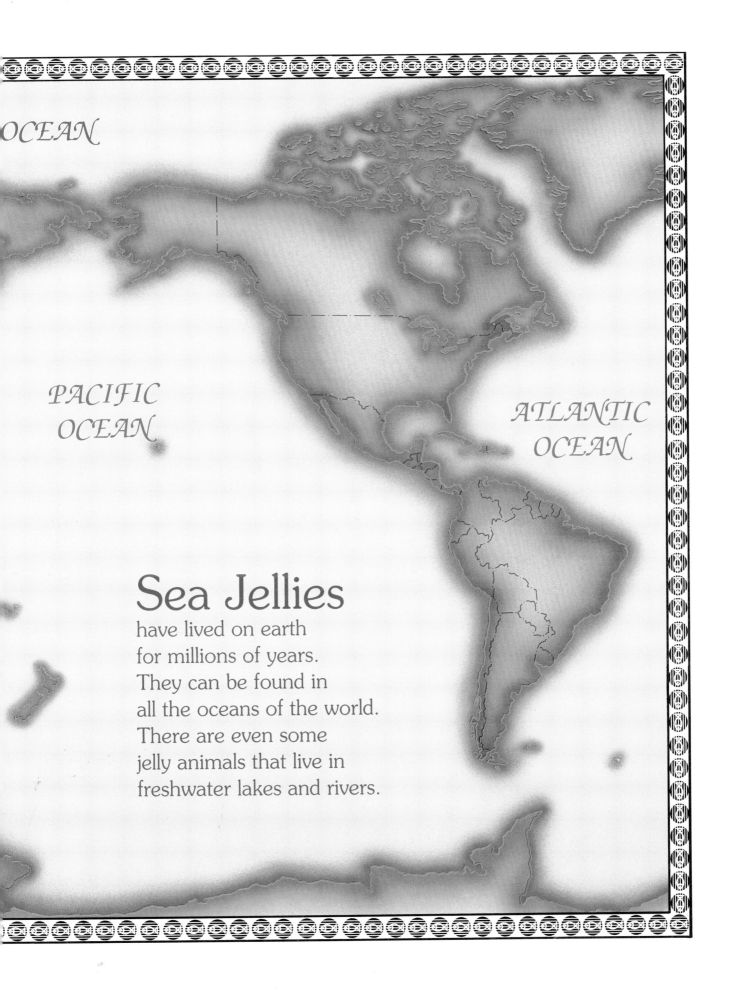

OCEAN

PACIFIC
OCEAN

ATLANTIC
OCEAN

# Sea Jellies

have lived on earth
for millions of years.
They can be found in
all the oceans of the world.
There are even some
jelly animals that live in
freshwater lakes and rivers.

# Sea Jellies

## Rainbows in the Sea

by Elizabeth Tayntor Gowell

A New England Aquarium Book

Franklin Watts
New York / Chicago / London / Toronto / Sydney

*For Emily*

## Acknowledgments

Many thanks to Dr. George Matsumoto and Dr. Larry Madin for taking time to share their vast knowledge of this topic and firsthand accounts of their field research experiences. Thanks also to the contributing photographers whose brilliant images make this book come alive. Ken Mallory and Lorna Greenberg provided valuable editorial guidance. Finally, my gratitude to Jay, Nina, John, and Sarah, without whose support this book would not have been possible.

Photographs copyright ©: Peter Arnold Inc.: frontispiece, pp. 6 right (both Fred Bavendam), 4 bottom (Norbert Wu), 7 bottom left (Sea Studios Inc.); Richard Herrmann: pp. 4 top, 34, 43; Herb Segars: pp. 6 left, 10 bottom left, 13 top right, 21; Norbert Wu: pp. 7 top and center left, 9 bottom, 11, 12, 13 left, 16, 18, 24 top, 32, 36, 44, 48, 49, 50, 51, 52; Animals Animals: pp. 7 bottom right, 10 right, 25, 28 top, 33 (all Oxford Scientific Films), 10 top left (Zig Leszczynski), 13 bottom right (Herb Segars), 19, 29 bottom (both Keith Gillett), 28 bottom, 29 top and center, 46 bottom right (all Peter Parks/OSF), 35 bottom, 38, 41, 42 (all Kathie Atkinson/OSF), 39, 40 (Fred Whitehead); Fred Bavendam: p. 8; New England Aquarium/Paul Erickson: pp. 9 top, 14; Lynn Cropp: p. 20; Ben Cropp: pp. 24 bottom, 27; Mike Johnson: p. 35 top; Peter Parks: p. 46 top and bottom left; L. P. Madin: p. 47; Ocean Images/Al Giddings: p. 53.
Diagrams copyright: © James Needham

Library of Congress Cataloging-in-Publication Data

Gowell, Elizabeth Tayntor.
    Sea jellies : rainbows in the sea / by Elizabeth Tayntor Gowell.
      p.    cm.
    "A New England Aquarium book."
    Includes bibliographical references (p.    ) and index.
    Summary: Introduces the anatomy, feeding, reproduction, and environment of various sea jellies throughout the world.
    ISBN 0-531-15259-6 (trade). — ISBN 0-531-11152-0 (lib. bdg.)
    1. Coelenterata—Juvenile literature. [1. Jellyfishes. 2. Coelenterates.] I. New England Aquarium Corporation. II. Title.
    QL375.6.G68   1993
    591'.5—dc20                    92-38515 CIP AC

J
593.7
9

# Sea Jellies

The frilly feeding arms of the purple-striped sea jelly (*Pelagia colorata*) may grow to be 20 feet (6.1 m) long. Divers search offshore in the open ocean to find these jelly giants.

It floats in the water like a spaceship from another world, then sinks slowly down into the sea. Deep purple bands mark its umbrella-shaped body. Eight long and slender pink tentacles hang down from the edges of its body and a frilly mass of pink and white, like ruffles on a dress, trails out from the center. As the animal sinks, it shimmers with rainbow colors in the sunlit water. A tiny fish swims by, too closely. It touches one of the tentacles and is instantly paralyzed—a victim of this strange and colorful creature of the ocean.

Sea jellies like this one are among the most beautiful and unusual animals on earth. Jellies are found in all the oceans of the world and even in some freshwater lakes and rivers. In the summer, you might see some washed up on a beach or floating in a harbor. And if you could peer into the icy waters of the Arctic or Antarctic oceans, you would see them there, too. Sea jellies have been discovered more than 3,200 feet (1,000 m) below the surface, and they probably exist even deeper.

Many sea jellies belong to a group of animals with the scientific name **Cnidaria** (from the Greek *knide*, meaning nettle, a plant with stinging thorns). Sea anemones, corals, hydroids, the Portuguese man-of-war, and the creatures commonly called "jellyfish" are all animals in this group. Scientists estimate that there are more than 9,000 species of Cnidaria. All of them share common characteristics:

jellylike bodies, tentacles, and stinging cells. But these animals are also so different from each other that scientists have split Cnidaria into three separate subgroups: the **Scyphozoans, Anthozoans,** and **Hydrozoans.** Some scientists think there should also be a fourth subgroup, the box-shaped **Cubozoans.**

The Scyphozoans are the most familiar jelly animals—free-swimming, bell-shaped animals best known as "jellyfish." Scientists prefer to call these jelly animals **medusae** (**medusa** in the singular), because they are not fish at all, and that is how we will refer to them in the rest of this book. In Greek myths, Medusa was a monster whose head was covered with writhing snakes instead of hair. The many tentacles surrounding the body of the sea jellies may have reminded early scientists of this mythical creature.

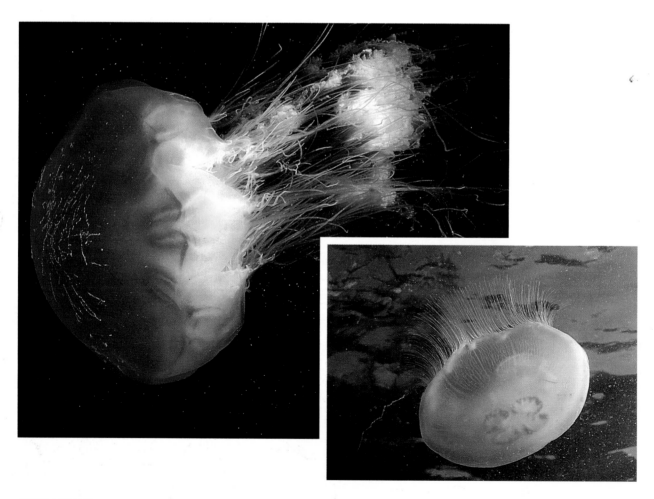

CNIDARIANS
Above: lion's mane jelly (left) and moon jelly (right). Facing page: top row, anemone, coral; middle row, hydroid, Portuguese man-of-war; bottom, Cubozoan.

The Anthozoans, or flower jellies, look like flowers, or upside-down bells on stalks when they are adults. The body of an adult flower jelly is called a **polyp.** Most polyps do not swim. With their adhesive base, they hold on to seaweeds or rocks and wait for food to come to them. Sea anemones, although polyps, are able to creep slowly along the seafloor or can sometimes detach themselves and then are carried along by currents. Coral polyps spend all their life in one place. Soft corals live on flexible stalks shaped like feathers, fans, or whips. Hard corals attach themselves to the ocean bottom with stony skeletons that protect their soft bodies. These hard corals live in colonies made up of hundreds, or even hundreds of thousands, of individuals. Through many generations of growth they have formed some of the largest natural structures on earth, such as the Great Barrier Reef off the coast of Australia.

In an emergency, some anemones can detach themselves, as the one shown here is doing to avoid becoming a sea star's dinner.

These soft coral animals from the Red Sea find some food the same way the hard, or stony, corals do—using their tiny arms or their tentacles to sting or capture their prey.

Most hard coral polyps are seen only at night, when they extend their tentacles to feed under the cover of darkness. During the daytime, or when disturbed, the polyps disappear into the safety of their hard, stonelike skeletons.

9

While the adult medusa jellies are free swimming, and the adult flower jellies are attached polyps, the adult jellies of the third Cnidarian subgroup—the Hydrozoans—can be either free swimming or polyps. **Hydroids** and freshwater **hydra** are examples of polyp Hydrozoans. Both hydroids and hydra form **colonies** in the shape of graceful feathers or tiny branching trees. The beautiful but dangerous Portuguese man-of-war is a special floating form of Hydrozoan.

This underwater garden of pink-hearted "flowers" is actually a colony of jelly animals. Despite their small size and delicate appearance, many hydroids can deliver a powerful sting.

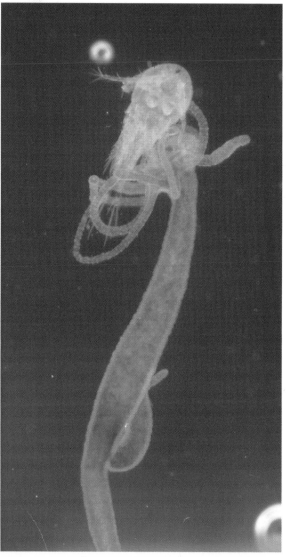

Hydroids have a swimming stage as well as a polyp stage in their life cycle. However, hydroid medusae are usually much smaller; some are only half the size of a thumbtack.

Hydra are freshwater hydrozoan jellies. This hydra's green color and shape provide camouflage. It hides among pond weeds and algae and waits for tiny insect larvae or plankton to drift into its tentacles.

**Ctenophora,** or the comb jellies, are another important group of animals related to the Cnidaria. Ctenophores exist in a variety of shapes. Some are called sea gooseberries, others sea walnuts because of their appearance. The name comb jelly comes from eight rows of

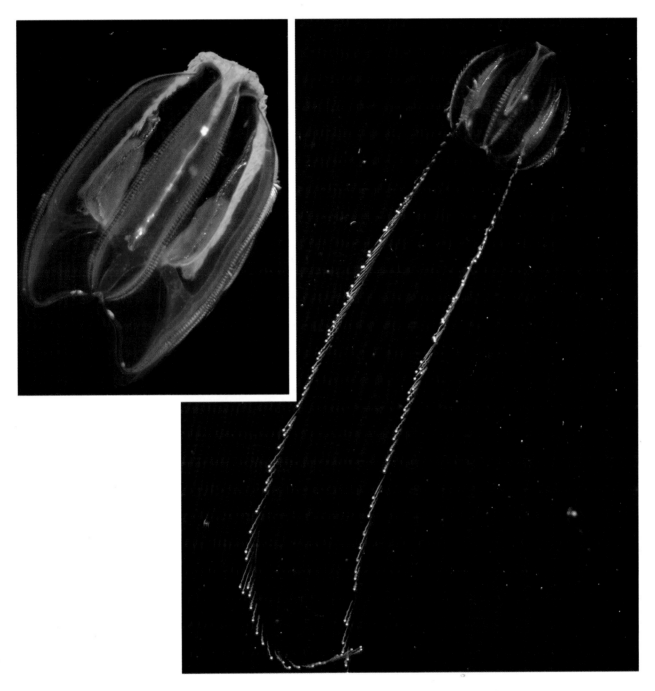

Some kinds of comb jellies have two long, sticky tentacles, which they first extend to trap food and then pull back into their stomachs. This feeding technique is like licking the sauce off a strand of spaghetti.

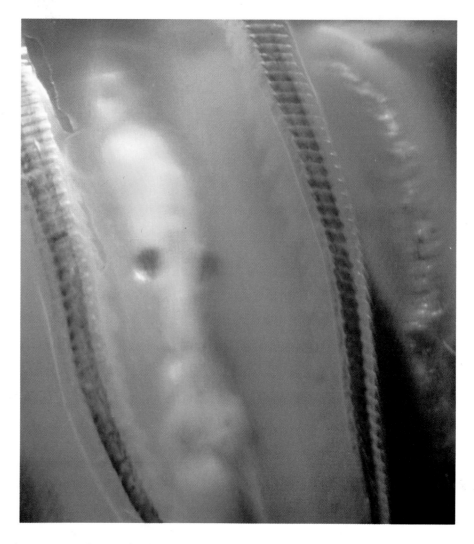

The beady-eyed creature in this comb jelly's stomach is a shrimplike crustacean called a krill.

tiny structures called cilia, which look like the teeth of a comb. Cilia are found on all comb jellies. The cilia constantly move back and forth to propel comb jellies through the water. As the cilia move, they bend light just as prisms do, making rainbow colors appear up and down the comb jellies' bodies.

Sea jellies come in all sizes. Imagine a jelly as wide as a backyard satellite dish. *Cyanea*, or the lion's mane jelly, is a real sea giant—as much as 12 feet (3.6 m) across, with tentacles 100 feet (30 m) long. Other jellies, such as the hydroid *Obelia*, are full grown at one-tenth of an inch (3 mm). The moon jelly *Aurelia* is usually 6 to 10 inches (10 to 25 cm) across, about the size of a dinner plate.

Pelagia

Lion's mane jelly

Tubularian hydroid

Sea jellies vary greatly in size and include the large *Pelagia* and the small polyps of the tubularian hydroid.

Medusa jellies were once called jellyfish, but they are not really fish at all. Fish are **vertebrates,** meaning they have backbones, and fish have gills and scales. The medusae and the rest of the sea jellies are **invertebrates**—they have no backbones. They also lack gills and scales. Many other ocean animals are invertebrates, including crabs, lobsters, snails, sea stars, squid, and sponges.

The feather star, crown-of-thorns sea star, and pajama nudibranch, or sea slug, are invertebrates, just as sea jellies are. They lack the backbone found in fishes, mammals, and all other vertebrates.

Sea jellies are very simple animals. They are 95 percent water, with a little bit of protein, fat, and salt. Most have a hollow body shaped like a bell or umbrella. The body is made of two thin layers of cells with a thick layer of jellylike material in between. The mouth is an opening at the center of the animal. Tentacles surround the edges of the bell or the umbrella. Some of the medusa jellies also have frilly arms around their mouths.

Sea jellies have lived on the earth for millions of years. Jelly fossils have been found in rocks 650 million years old off the coasts of England, South Africa, and Australia. Scientists even found one very ancient fossil that looks like a sea jelly in rock layers deep within the Grand Canyon. Paleontologists (scientists who study fossils) believe that the fossils were formed when dead jellies sank to the soft mud bottom of an ancient sea and were covered with fine layers of mud and silt. The fossils we see today are impressions left by the jelly body.

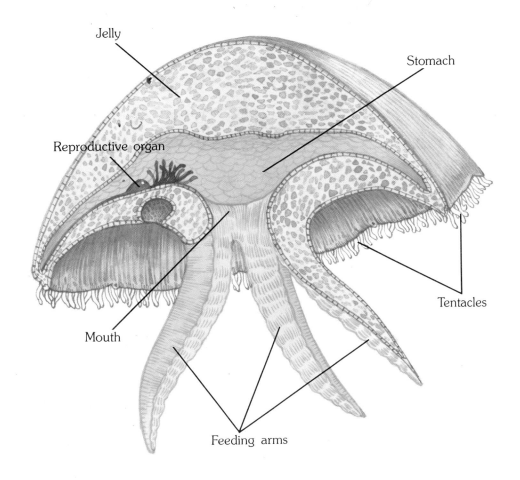

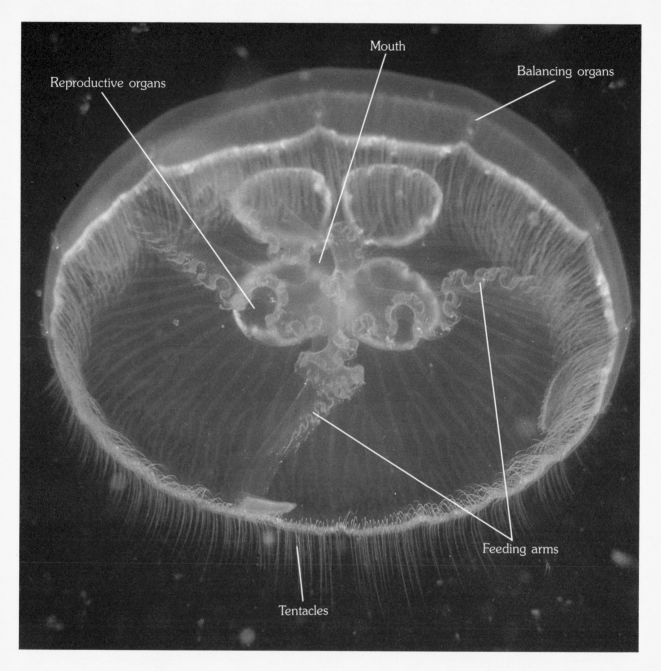

Reproductive organs

Mouth

Balancing organs

Feeding arms

Tentacles

A view of the underside of a moon jelly shows many of its survival secrets. The white dots along the edges of the body contain light- and motion-sensitive cells that help the jelly to detect sunlight and sense whether it is upside down or turned sideways. A dense ring of tentacles and frilly oral arms enable the jelly to catch and feed on tiny animals.

# How Jellies Work

Sea jellies are among the simplest animals in the world. They do not have many of the organs that you or most complex animals have. They have no brain, no liver, no kidneys, no heart, no lungs, or gills. They do not need blood because oxygen and nutrients don't have to be transported long distances to their body cells. Jellies absorb oxygen directly from the water. It passes through the thin walls of their body tissues.

The mouth of the medusa jelly is located on the underside of its body, in the center. It is the only opening into and out of the medusa's stomach, a hollow cavity where food is digested. Like oxygen, nutrients pass directly into the cells surrounding the stomach and through the jelly layer to other parts of the body. The frilly arms some medusae have around their mouths help transfer food to the stomach. Once food is digested, wastes are passed back out through the mouth.

## Getting Around

A medusa swims by contracting muscles around the edge of its body. As the muscles contract, water is pushed out of the hollow body and the medusa is jet-propelled in the opposite direction. Squids and octopuses use the same technique for making fast getaways. Although this may not seem like much of a swimming technique to us, a medusa can travel long distances just by pulsing its body. One species of

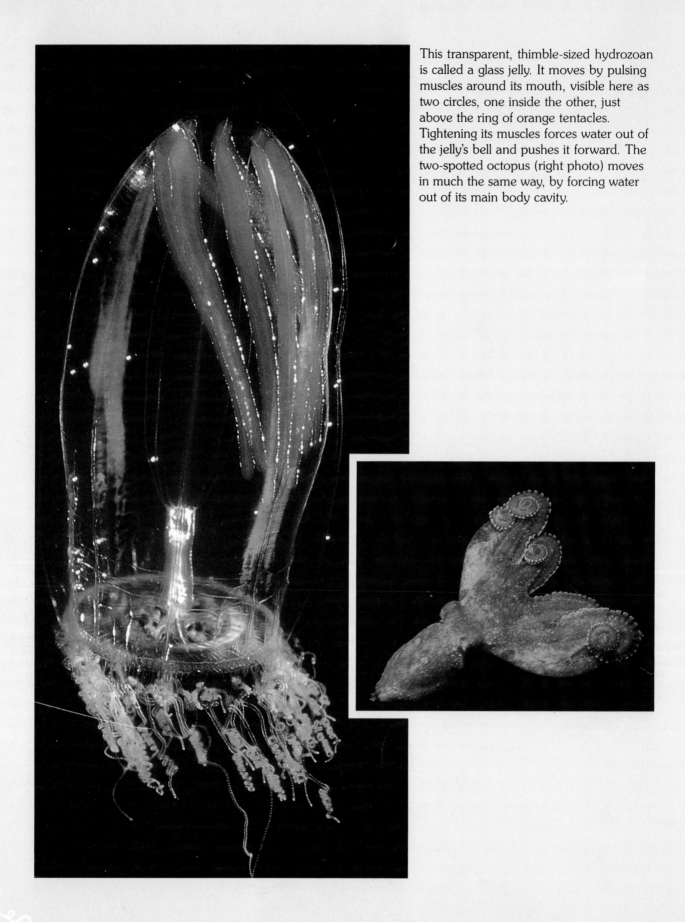

This transparent, thimble-sized hydrozoan is called a glass jelly. It moves by pulsing muscles around its mouth, visible here as two circles, one inside the other, just above the ring of orange tentacles. Tightening its muscles forces water out of the jelly's bell and pushes it forward. The two-spotted octopus (right photo) moves in much the same way, by forcing water out of its main body cavity.

Mediterranean medusa, *Solmissus albescens*, is only about one and one-half inches (3.75 cm) across. It swims up and down through the **water column** a distance of 3,600 feet (1,100 m) a day, following the tiny sea animals, or **zooplankton,** on which it feeds. This distance would be equal to a 33-mile (53-km) swim for a 6-foot (1.8-m) person.

Although medusae can move very well up and down in the water column, their swimming technique is no match for strong currents, wind, and waves. At certain times of year, beaches are littered with the bodies of medusae washed up by storms or strong seasonal winds. Once on shore, the medusae have no way to get back to the sea. Their jellylike bodies dry up, and only filmy circles are left behind on the sand. Some close relatives of medusa, such as the Portuguese man-of-war, have floats to keep them at the ocean's surface. These animals don't really swim, but drift wherever they are taken by the wind.

Some kinds of jellies travel great distances and are surprisingly strong swimmers. However, ocean winds, currents, and tides are sometimes more powerful than the jellies, and the animals are washed up on shore.

# How Do Jellies Sense Their World?

Do sea jellies feel, hear, smell, taste, and see their undersea world? Unlike more complex animals, their senses are limited. Sea jellies have touch receptors on their tentacles and around their mouths to help capture food. These touch receptors may also detect vibrations in the water caused by the movement of a fish, crab, or other animal swimming by.

Sea jellies do not have a nose or tongue. They have special cells that smell and taste scattered all over their bodies. Sea jellies do not have eyes like human eyes, but many have light-sensitive organs around the margins of their bodies. In most cases, these organs do not detect shapes or movement, but allow the jelly to tell light from dark. Jellies can tell up from down by sensing the sunlight at the surface of the ocean.

Sea jellies use their balancing organs to orient themselves up, down, or sideways in the water.

Light-sensitive organs at the edges of this lion's mane bell give the jelly
information that allows it to move toward or away from sunlight.

Sea jellies also stay upright with the help of special balancing
organs, which are located along the outside edges of their bodies.
These fluid-filled sacs each contain a tiny, stonelike granule. When the
sea jelly tilts too much to one side, the granule touches cilia, which
stimulate nerve endings. These nerves cause the jelly's muscles to
contract, putting it back on course. We have tiny stones in our middle
ears that move around as our heads tilt, sending signals to the brain
that help us keep our balance.

# The Jelly Life Cycle

Most young sea jellies look nothing like their parents. Just as a frog must spend time as a tadpole before it grows up, sea jellies go through several life stages before they reach adulthood.

Let's look at the moon jelly, *Aurelia*, as an example. An adult moon jelly has a classic, bell-shaped medusa form. Some moon jellies are male. Some are female. The females produce eggs and hold them on the frilly arms around their mouths. The male medusae produce sperm and release them into the water. The females then take in the released sperm to fertilize their eggs. Each fertilized egg stays on the female jelly's mouth arms, where it grows into a tiny, pancake-shaped **larva.** The female moon jellies then release their larvae to the ocean, where they ride the currents and waves until they reach a rock or other hard surface on which they settle. Unlike their floating, free-swimming parents, the baby jellies attach themselves in one place, and stay there.

Once settled, each larva transforms itself into a polyp. The polyp, which at first looks like a tiny sea anemone, uses tiny tentacles to capture zooplankton for food. As it grows, the polyp may produce buds, or branches, that eventually break off and become separate polyps. These budded polyps are entirely new animals.

After several months, or sometimes several years, the polyps start to undergo an amazing transformation. Grooves form in the body of each polyp. The grooves become deeper and deeper until they go all the way through the polyp's body, creating a stack of tiny disks. Each of these disks is a baby moon jelly. These tiny jellies, each only 1/8 inch (0.3 cm) across, gradually break off from the stack, one by one, and swim away on their own. Those that survive become the next generation of adult moon jellies.

From each larva that survives to reproduce, many moon jellies eventually develop—some from new polyps that bud off from the original polyp, and others from the original polyp itself.

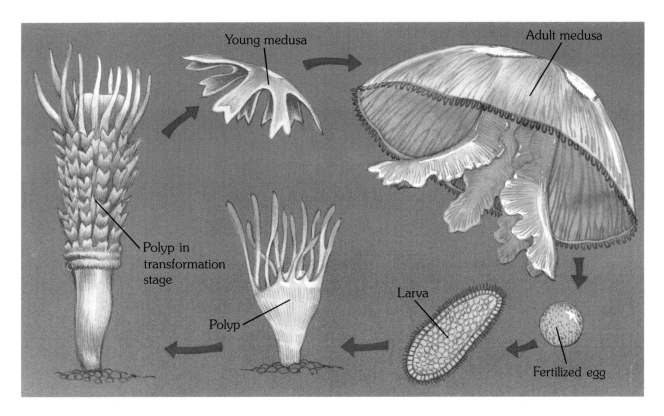

The Life Cycle of a Scyphozoan

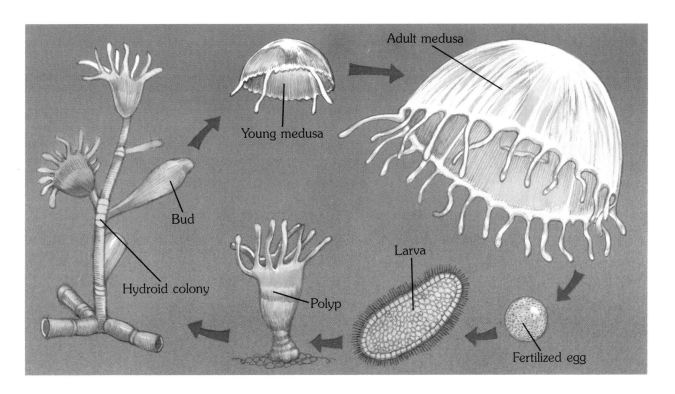

The Life Cycle of a Hydrozoan

The clover-leaf-shaped reproductive organs of the moon jelly produce eggs in the female and sperm in the male. Once a female's eggs are released, they are held on the feeding arms, where sperm from a male jelly then fertilizes them.

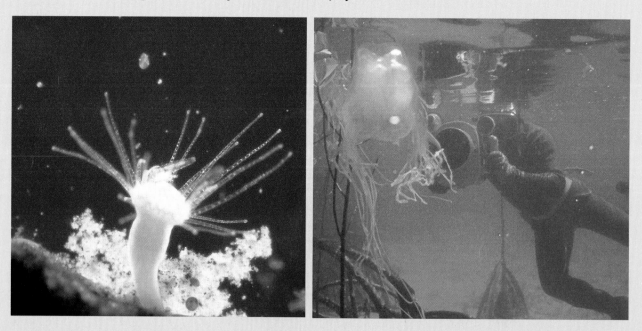

Fertilized sea-jelly eggs produce pancake-shaped larvae, which settle on a firm surface and grow into polyps. This sea-wasp polyp gets its food the way hydroids and anemones do: by capturing it with its stinging tentacles. As the polyp matures, it changes into a free-swimming medusa and continues through the life cycle.

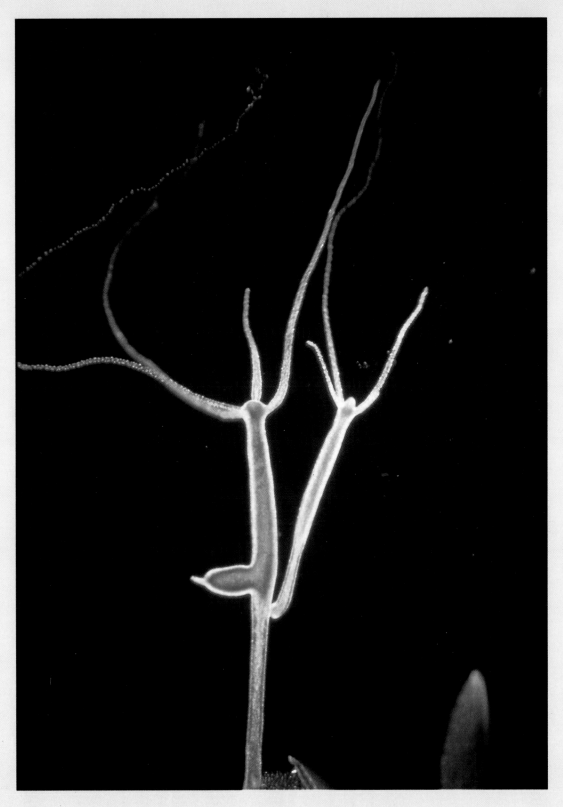

Jellies have a second way to reproduce themselves—through budding.
In this photograph a hydra has formed an exact copy of itself,
which is budding off to become a new jelly.

# Stinging

*"I do not care to share the seas*
*With jellyfishes such as these,*
*Particularly. . . . Portuguese."*

The Bestiary of Flanders and Swann

Jelly animals are beautiful to look at with their colors of iridescent pink, purple, and red, and their flowing tentacles. But most people don't want to get too close—many sea jellies sting!

Some jelly stings are so mild you might not feel them. Some, however, are painful and can cause sharp, burning sensations and red, swollen welts. A few can even cause death.

The most deadly jelly animal is the sea wasp. This creature lives in tropical waters throughout the world. In the southeast Pacific and Indian oceans, one species, *Chironex fleckeri*, is believed to have caused at least sixty deaths over the past hundred years.

In American waters, the most dangerous species of sea wasp is *Chiropsalmus*, which is seen in the Gulf of Mexico and off the southeastern shore of the United States. A sting from this jelly can make it hard to breathe, and the victim might need to be taken to a hospital for treatment. But *Chiropsalmus* is not a killer like the Australian sea wasp. We have no record of any death ever caused by a sea wasp along the coastline of the United States.

## The Power of Nematocysts

How can a blob of living jelly about the size of a coconut be so dangerous? The stinging power of jellies comes from special cells on their tentacles and other body surfaces that contain capsules called

Tropical swimmers beware! Though not the largest of the sea jellies, the sea wasp is the most deadly.

Top: The tentacles of the Portuguese man-of-war contain some of the most powerful stinging cells in the sea. Bottom: A close-up view of the tentacles.
Facing page: Through high magnification, these photos show nematocyst capsules before (top) and after some (center) are discharged. Bottom: Triggered by touch or a chemical cue, a long hollow tube is thrust out from each capsule. If the tube pierces a swimmer's skin, a painful toxin is injected.

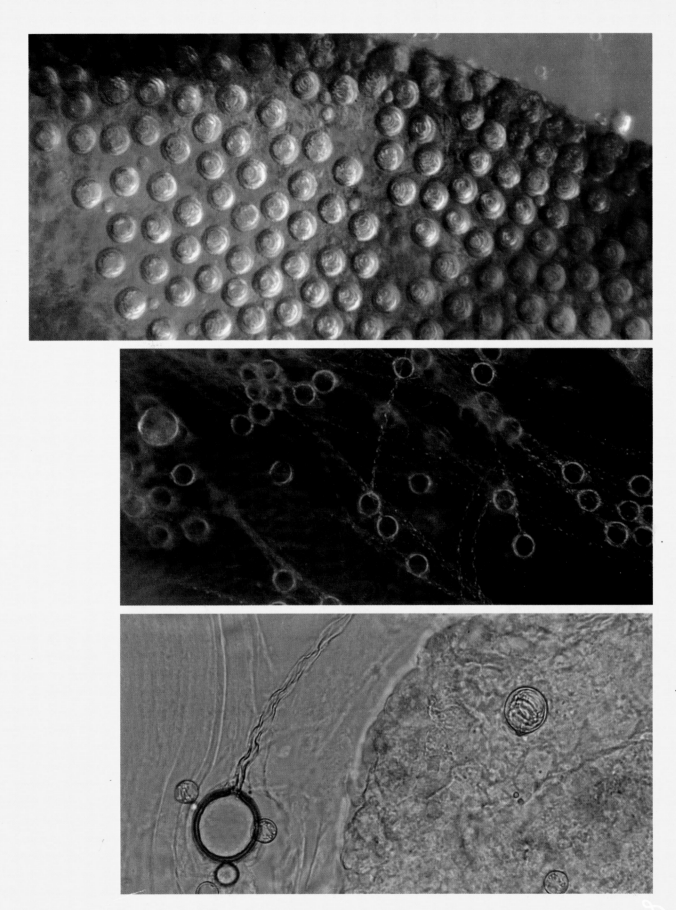

**nematocysts.** Many nematocysts have trapdoors with tiny hairlike triggers. Coiled inside the capsule is a long, hollow tube. When a swimmer's leg touches the trigger, the trapdoor opens and the tube is everted, meaning it is turned inside out, as it extends from its capsule. The whole event takes place in milliseconds.

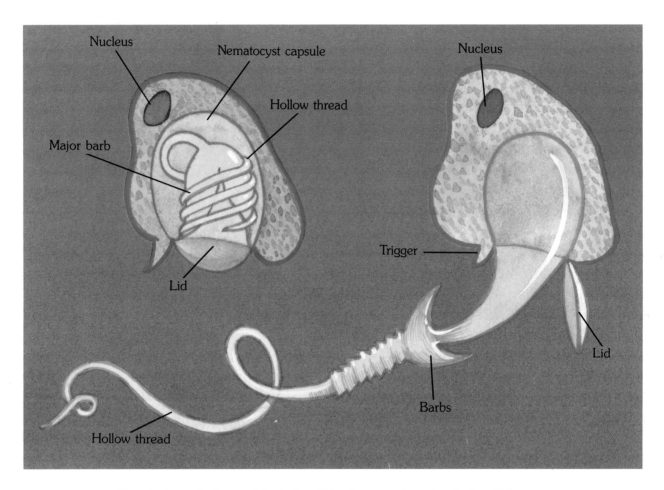

The stinging cell of a sea jelly before firing the nematocyst and after firing

The nematocyst tubes of some stinging jellies work like the hypodermic needles doctors use to give injections. When the tube pierces the victim's skin, a paralyzing toxin is injected. Other jellies have tubes with dozens of tiny hooks that stick onto their prey like Velcro. Some even work like tiny lassos that wrap around and trap tiny animals. Scientists have found 30 different types of nematocyst tubes—some with barbs, some smooth, some short, and some 300 times as long as the nematocyst capsule is wide.

# Prevention and Treatment

How do you avoid being stung by a jelly? First, get to know the sea life in your area. Stay away from beaches when large numbers of stinging jellies are blown ashore by high winds. After a storm, look around to see if stinging jellies are in the water. Tentacles torn off an injured jelly can still sting. A dead jelly washed up on a beach should be avoided. The stinging power of nematocysts can remain even after the jelly is dead.

In Australia, lifeguards used to wear panty hose to the beach—one pair on their legs and one pair, with holes cut in it to fit over their heads and hands, on their upper body. They might have looked funny, but the fabric protected them from the short but deadly nematocysts of the sea wasp. Today, Lycra bodysuits have been designed for the same purpose. Some swimmers also coat their bodies with Vaseline or lanolin to prevent sea jelly stings.

What do you do if you are stung? First, get out of the water. Researchers believe that in a few cases, good swimmers have drowned because of jelly stings. Once on land, try to remove the tentacles, being careful not to get stung again. Do not try to rinse the tentacles off by washing the skin with water. This could cause more nematocysts to fire. Do not brush or rub the tentacles off. This will cause nematocysts to fire, too. Some people have severe allergic reactions to even mild stings, so it is best to see a doctor.

If you were swimming off the north Atlantic coast of the United States and were stung, the stinger was probably *Cyanea*—the lion's mane jelly; in the Chesapeake Bay area, it was *Chrysaora*—the sea nettle. South of Virginia Beach, along the Gulf of Mexico, on the west coast of the United States, or in Hawaii, the stinger was probably *Physalia*—the Portuguese man-of-war.

# What's for Dinner?

Why do jellies have such a powerful sting? The stinging cells are weapons that enable the jellies to catch their food. Jellies eat almost

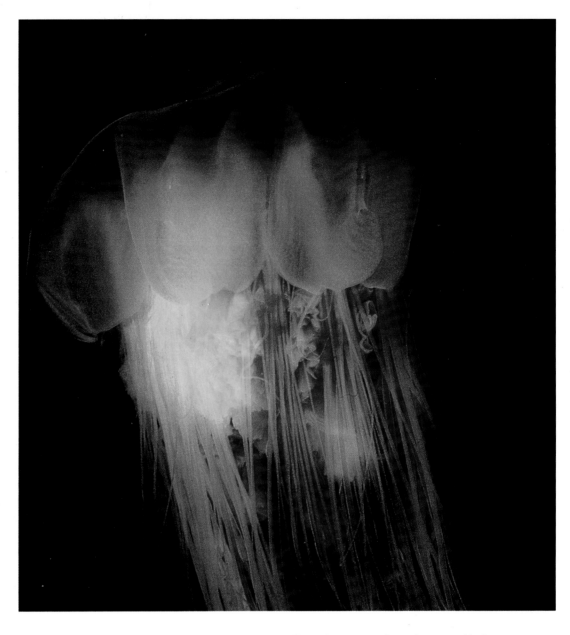

If you are stung while swimming in Chesapeake Bay, the sea nettle is the most likely cause.

anything they can capture, from tiny floating animals called zooplankton to fish and even other jellies. Most jellies fish for their food with tentacles that may be many times longer than their bodies. The Portuguese man-of-war has tentacles up to 60 feet (18 m) long. When an unlucky fish or tiny crab touches one of these, the nematocysts shoot out and the victim is trapped. The man-of-war then contracts its tentacles, reeling in the captured prey. After a nematocyst is fired, the jelly's body produces a new one to replace it.

The Portuguese man-of-war's tentacles are deadly to its prey, like this small fish.
When they are not extended for trapping food, the tentacles rest—coiled
like springs—close to the jelly's body.

## Who Eats Sea Jellies?

The stinging cells of sea jellies help them eat, and also help keep them from being eaten. Surprisingly, there are a number of ocean animals that find jellies, even the big stingers, tasty prey. The ocean sunfish sucks down jellies with a noisy slurp. And it must eat a lot of them, since these fish can grow to a whopping 1,500 pounds (680 kg) eating almost nothing but jellies.

Some fish can get past a sea jelly's defenses. An orange garibaldi fish nibbles on the tentacles of a purple-striped sea jelly.

Sea turtles also find jellies a tasty treat. Loggerhead sea turtles are especially fond of the Portuguese man-of-war. When winds or currents bring huge numbers of these jellies together in one part of the ocean, sea turtles gather and feast on the purple floats. They brush away the stinging tentacles with their flippers between bites, but their sensitive eyes still swell from the stings.

One jelly predator is a small blue sea slug, or nudibranch, called *Glaucus*. This animal lives off the coast of Australia and feeds on *Physalia* and other sea jelly tentacles. When swimmers reported that they had been stung by this little nudibranch, scientists investigated and discovered unexploded man-o-war nematocysts in the nudibranch's tissues. It appears that the nudibranch can move stinging capsules from the man-of-war into its own body, thus equipping itself with stolen weapons.

Above: While a sea jelly's stinging cells help it capture food, they don't always stop other animals from making jellies their dinner. Here an ocean sunfish is about to swallow a by-the-wind sailor (*Velella velella*). Look closely to see the tiny jelly floating in front of the sunfish's pink mouth.

Below: Who is eating whom in this mass of jelly bodies floating at the sea's surface? If you look carefully you should find by-the-wind sailors, Portuguese man-of-war jellies, and frilly-limbed nudibranchs, or sea slugs. Remarkably, these sea slugs can eat man-of-war tentacles without being harmed. They even take the stinging cells and use them for their own defense.

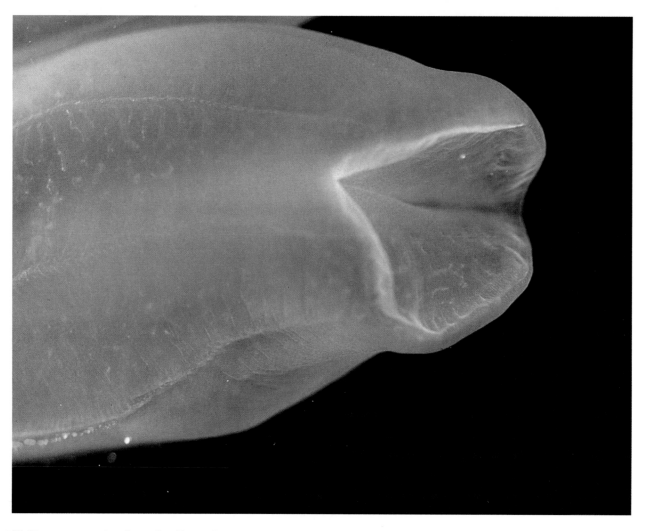

Unlike most comb jellies, this *Beroë* has no tentacles to help it feed. With its wide mouth it "vacuums" in other kinds of sea jellies, some as big as itself.

Some jellies even eat other jellies. One jelly predator can grow to be as much as a foot (30.5 cm) long. It is the comb jelly *Beroë*. Like other comb jellies, *Beroë* does not sting. It does not even have tentacles. Basically, this jelly is a swimming mouth that opens wide to vacuum in its prey, usually smaller species of comb jellies. Inside its mouth, *Beroë* has specialized structures that are shaped like fangs or the teeth of a saw. These are used to trap or take bites out of its prey.

In some parts of the world, sea jellies are enjoyed as a delicacy. In Japan, China, and Korea, people eat salted and dried jellies that look like thin white pancakes and are chewy, like rubber bands. Other types of jellies are soaked in salt water, steamed, then seasoned with spices and served.

# Some Unusual Jelly Animals

## *Physalia*, the Portuguese Man-of-War

*Physalia* is one of the most beautiful and most feared of all the jellies. Most jellies must swim to stay at the ocean's surface. Not the Portuguese man-of-war. This unusual animal has a gas-filled float, or balloon, that lets it ride high on the ocean waves. The gas is produced by cells in the jelly's body. The floats are a lovely pinkish purple or blue color, and are shaped somewhat like the helmets of early Portuguese conquistadors, hence the name Portuguese man-of-war.

Surprisingly, each Portuguese man-of-war is not just one creature, but a colony of hundreds of units that live and work together. The float is a single structure that is specialized for keeping the colony afloat, but it cannot capture food. Hanging down from the float are structures containing many specialized polyps and tentacles, all with tasks that they perform for the colony. The hunters of the colony are polyps that are specialized for stinging and capturing prey. Other polyps have the job of stomach—they digest the food for all the colony members. Finally, some members are in charge of reproduction for the colony.

The beautiful purple float of the Portuguese man-of-war is like a living balloon. To keep its float from drying out, the jelly wets it several times a day by tipping to one side, then the other.

Sometimes great numbers of *Physalia* are blown ashore, making swimming very dangerous. The float of the Portuguese man-of-war can be a foot (30.5 cm) long and is easily seen at the ocean's surface. But even swimmers who carefully stay away from *Physalia*'s float may be victims of its very long tentacles, which may trail up to 60 feet (18 m) behind.

The nematocysts of the man-of-war are so powerful that the barbs can pierce a thin rubber glove. And the sting of this jelly is very painful, sometimes even deadly. One scientist said, "A thimble-full of venom from a Portuguese man-of-war is enough to kill a thousand mice." People, too, have died after being stung by the man-of-war. Some scientists think the victims may have had allergic reactions to the venom of *Physalia*, similar to the reaction that some people have to bee stings. Many other people have been stung quite badly and survived, though the pain from the jelly's stings lasted for several days.

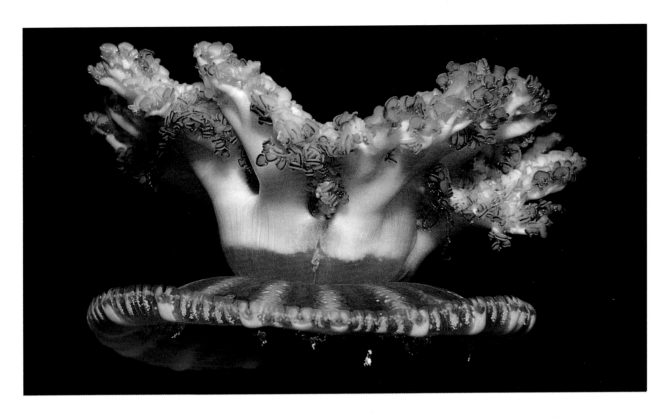

The upside-down sea jelly, *Cassiopeia*, like the coral polyp, catches some of its food. However, both also rely on nutrients produced by plants living in their tissues.

## *Cassiopeia*, the Upside-Down Jelly

Most medusa jellies float in the water, with their bell-shaped bodies up and their mouths below; but not *Cassiopeia*, the upside-down jelly. In the warm, shallow waters of mangrove islands in Florida or the Caribbean Sea, you'll find these unusual jellies upside down in the mud. They contract their bodies gently, not to swim for the surface but to hold themselves flat against the bottom.

*Cassiopeia* are also called farmers because they have microscopic plants living inside their bodies. These plants provide the jelly with oxygen and food through the process of **photosynthesis.** The jellies, in turn, provide safe homes for the plants that use carbon dioxide and other jelly wastes to grow. When scientists find two different species of plants or animals living together, they call the relationship **symbiosis.** The golden green color of *Cassiopeia* comes from the many tiny plants living in its body.

*Cassiopeia* hold themselves flat against the floor of the sea.

Though it looks like most medusa jellies, the upside-down jelly does not have true tentacles. The frilly lobes on the underside of its body are actually extensions of its main mouth tube. And each of these extensions has a tiny mouth of its own for eating tiny prey. Despite these many mouths, the food that grows inside *Cassiopeia* may be more important than the food it catches. If these jellies are kept in laboratory tanks where their plants do not get enough sunlight, the jellies shrink. Once placed back in the sun, the animals start to grow again.

## *Velella*, the By-the-Wind Sailor

The little, purple jelly animal *Velella* looks more like a miniature sailboat than a sea jelly. The name *Velella*, which means "little sail," comes from the animal's unusual shape. *Velella* has a flat, thin, oval body with a triangular projection on the top that looks like a sail and serves the same purpose. Even in a light wind, *Velella* moves briskly

The by-the-wind sailor, *Velella velella*, is pushed along the ocean's surface by the wind.

*Velella*'s free-ranging life-style doesn't protect it from predators such as the *Janthina* snail, seen hanging from its raft of bubbles while it nibbles on the jelly.

over the waves, swept along by the breeze on its sail. To keep its sail at the ocean's surface, *Velella* has an internal float made of tiny tubes filled with gas. At certain times of year, thousands of these tiny sailors may be blown ashore. As their thin, purple bodies dry, glasslike "skeletons" of the floats are left behind.

*Velella* is more closely related to the Portuguese man-of-war than to the medusa jellies. This little sea jelly grows to be only about 2 inches (5 cm) long and its short tentacles are not dangerous to humans. It feeds on tiny prey—copepods, fish eggs, and other zooplankton.

An animal that is sometimes found with *Velella* is the violet snail, *Janthina*. The snail is a **parasite** and feeds on *Velella*. To get around on the ocean's surface, *Janthina* blows a raft of tiny bubbles that floats the snail to its prey.

# Salps—Weird and Wonderful Colonial Jellies

Some of the ocean's strangest jelly animals are not Cnidarians. They are more closely related to human beings than to jellies. These weird jelly animals are salps, oceangoing members of the scientific group of animals called sea squirts. Most salps are transparent, and look like simple animals, but unlike other jellies they are very complex. Sea squirt larvae have a primitive spinal cord that makes them cousins of vertebrates, though they never develop a true backbone.

Salps eat phytoplankton—tiny, floating ocean plants. They suck water into their mouths by contracting circular muscles located all around their barrel-shaped bodies. A mucous net inside the salp's body filters tiny plants from the water. The water is then released through an opening at the other end of the animal. This squirting action jet-propels the salp forward.

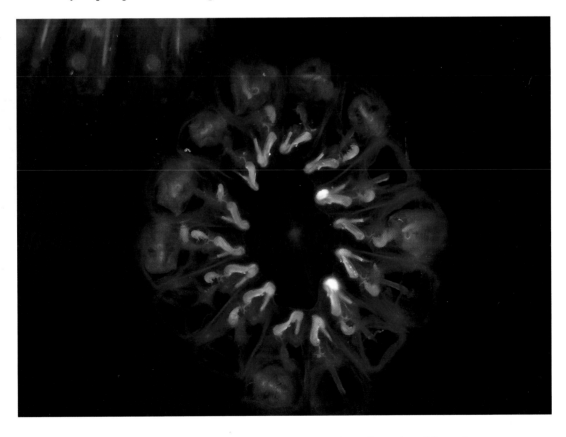

The unusual jellies called salps have primitive spinal cords early in their life cycle. The mature animals, however, are invertebrates, animals without backbones, like other jellies of the sea. Salps live as individual animals or in chains of many linked animals.

Like other jellies, salps are found in both individual and colonial forms. Some beautiful colonial salps are mostly transparent with a few brightly colored marks that look like stars. These salp colonies are made up of one parent salp along with many new salps that budded off from the first. Budding salps often form chains that can be 6 feet (1.8 m) long or longer, made up of dozens of individuals. In motion, a giant salp chain twists and twirls like a great ocean snake.

A salp colony

# Deep-Sea Jellies

Among the most beautiful jelly animals are those people usually see only in photographs. They live in the cold, dark, deep sea—hundreds, even thousands, of feet beneath the surface. Scientists once thought that the deep sea was like a watery desert with little animal life. But that idea is changing. With new ways to explore the ocean's depths, we are finding that it is much richer in life of all shapes and sizes than we ever expected.

The deep sea is still a new frontier. Before the invention of SCUBA (Self Contained Underwater Breathing Apparatus) in the 1940s, scientists studying marine life could not venture underwater for any length of time. By taking tanks of air with them, the observers could stay underwater for up to an hour or more. But the depth of these explorations was limited to 100 or 200 feet (30 or 61 m). Today, small submarines, specially designed for deep-sea research, take scientists down to depths of 20,000 feet (6,100 m), although depths of 2,000 to 3,000 feet (610 to 914 m) are more common. Here researchers can explore an undersea wilderness about which very little is known.

One surprise for these deep-sea explorers was the number of jelly animals that they found—medusae, anemones, and corals of fantastic colors and sizes. In fact, scientists now believe that jelly animals may be one of the most common types of animal life in the ocean depths.

Scientists have also found that the deep sea is not always dark. Some deep-sea jellies can turn on bright lights. Like underwater lightning bugs, they give off sparks of blue, green, or blue-green. This living light, called **bioluminescence,** is caused by a chemical reaction inside the jellies. Scientists are not sure why jellies glow in the dark. It

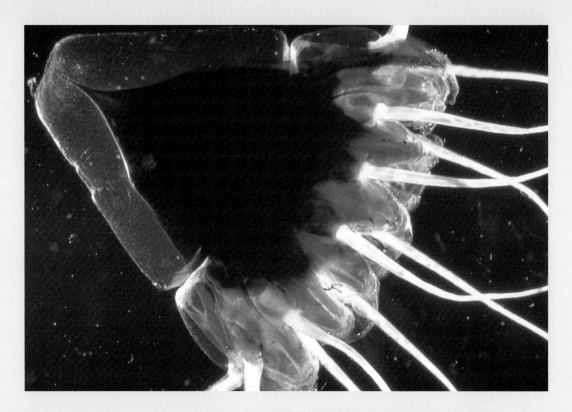

The colors and markings of sea jellies can be as varied as the patterns created by a kaleidoscope. Some of these jewel-like jellies were photographed 6,000 feet (1,800 m) below the ocean surface. As deep-sea exploration continues, scientists expect to discover many new species.

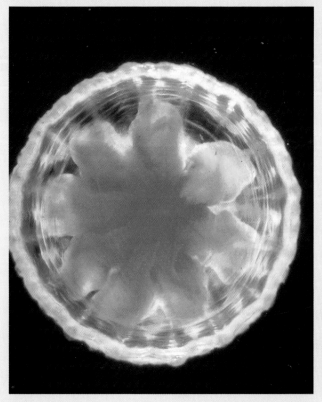

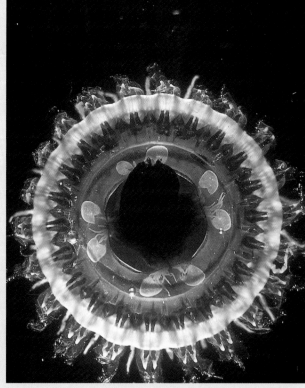

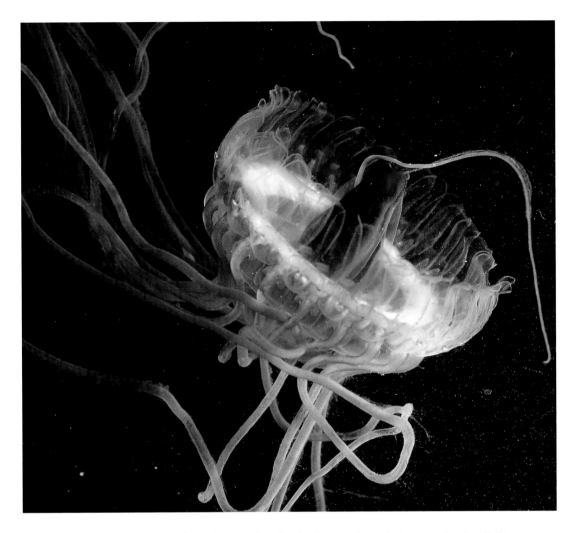

Many deep-water jellies, such as this comb jelly *Atolla*, produce their own glowing light. This is caused by a chemical reaction in their bodies.

may be to startle or confuse predators, or to attract their tiny zoo-plankton prey. Many sea jellies that live on the surface also produce bioluminescence. Sea captains sometimes see the ocean surface glow at night with the light of hundreds of jellies disturbed by a passing ship.

Most deep-sea jellies are brilliantly colored. Purple and red are the most common shades. The reason for these brilliant colors is a mystery. For jellies that live at the surface, scientists know that pale or transparent colors act as camouflage, making jellies almost invisible to predators. But in the perpetual darkness of the deep sea, what role can bright colors play?

Scientists are constantly developing better techniques to collect jelly animals, in hopes of expanding their understanding of how these rainbows of the sea are adapted to life in the ocean. Here at a depth of about 60 feet (18 m), a diver collects jellies and other marine life for examination in the laboratory.

Perhaps the color is simply a beautiful by-product of what these jellies eat. Or it may be an adaptation to hide what the jellies have eaten. Many of the tiny crabs, shrimp, and fish that deep-sea jellies eat also glow in the dark. If the jelly predators were transparent, their last meal would glow inside them like a giant underwater light bulb, making them easy targets for a hungry fish or squid. Instead, their dark body colors hide the bright lights of the animals they eat. A recently discovered deep-sea species of comb jelly, *Lampactena sanguineventer*, is called "bloody belly jelly" because of its dark red stomach.

Although we are learning more about the deep sea, there are still limits to studying life in the ocean depths. First, scientists can go only as deep and as far as the research submarines that carry them. Second, on very deep dives, they cannot leave the vessels. Therefore, much of what we know about deep-sea life is based on photographs and accounts of what scientists have seen from the windows of submersibles. Finally, although some research subs are equipped with mechanical arms and baskets for collecting scientific specimens, it is very difficult to maneuver this equipment to catch a moving animal. And jellies are very fragile—when touched, they break apart or are injured, making it difficult to bring them back alive and in one piece.

Sea-jelly scientists constantly look for new ways to learn about marine life. Here a green dye is released around a sea jelly to reveal the water currents produced by the swimming jelly. This information will be used in studying animal locomotion.

# The Future of Sea Jellies

For many years, jelly researchers were more interested in finding ways to get rid of these fascinating animals than they were in understanding them. Today, scientists appreciate the amazing things these simple creatures can do and are beginning to recognize the important role they play in the chain of life in the oceans.

What kinds of research do jelly scientists do? Some are continuing to search for an effective antivenin that will save victims of sea wasp

As the science of studying sea jellies advances, we will learn more about how jellies survive and how other animals live alongside them in the sea. Some jellies are like floating apartment houses. Small fish or invertebrates hide among their tentacles, safe from other predators. Scientists once believed these animals were immune to the jelly's sting. Now they believe the tiny fish are just very good at avoiding contact with the stinging tentacles.

stings. Others are studying the chemicals in medusae and other jellies for possible use in treating cancer and other diseases. One of the bioluminescent chemicals found in a medusa jelly from the Pacific Northwest has already been found to be useful in certain types of medical research. This substance allows doctors to trace the movement of specific chemicals through the body.

Some researchers are discovering and collecting jelly animals that are new to science. Others are researching how these animals live— what they eat, how they reproduce, and what special features they have to survive. This may help explain why jellies may appear in swarms at certain times of the year and offer clues about how to control them.

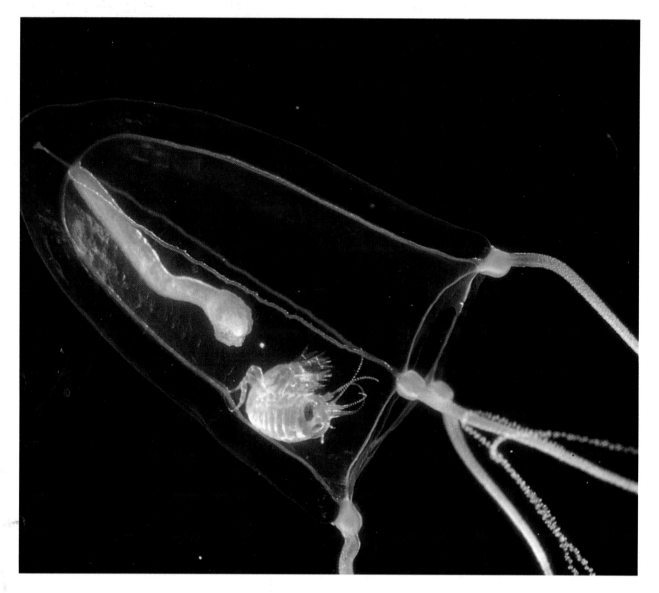

The tiny marine animal called an isopod steals both a ride and a meal from this sea jelly in the Arctic. The isopod is a parasite, taking food from its host animal without giving anything in return.

One of the early challenges in jelly research was to develop laboratory equipment and techniques for keeping these unusual animals alive. Since it was too difficult to collect enough of the tiny animals jellies eat, scientists learned to grow food for the jellies. They also designed special round tanks, with no flat walls for fragile jelly animals to bump into. Now that these techniques have been perfected, many more types of jellies can be collected and studied. As this research continues scientists are sure to discover many more amazing things about these animals and the ocean world in which they live.

# Glossary

**Anemones** (uh-NEH-muh-nees)—Jelly animals closely related to medusa jellies. Anemones are often brightly colored and have a polyp form that makes these animals look like underwater flowers. Most anemones live alone rather than in colonies.

**Anthozoans** (an-tho-ZO-uns)—A subgroup within Cnidaria made up of sea anemones, corals, and other jelly animals with a polyp form.

**Bioluminescence** (by-oh-LOO-muh-NES-ens)—A chemical process in living organisms that produces light.

**Cnidaria** (ni-DA-ree-uh)—The scientific group to which medusa jellies, sea anemones, corals, and related animals belong. This group of animals was formerly called Coelenterata.

**Colony**—A group of animals, all of the same kind, that live together in a way that benefits each member of the group.

**Ctenophora** (teen-uh-FOUR-uh)—The scientific group to which comb jellies belong.

**Cubozoans** (kyu-bo-ZO-uns)—The name proposed by some scientists for a fourth Cnidarian group, which would include box-shaped jellies such as the sea wasp.

**Hydra** (HI-druh)—Small jelly animals with a polyp form that are found only in fresh water. Hydra form colonies shaped like feathers or tiny trees.

**Hydroids** (HI-droid)—Small jelly animals with a polyp form found only in salt water. Hydroids form colonies shaped like feathers or tiny trees.

**Hydrozoans** (hi-druh-ZO-uns)—A subgroup within Cnidaria made up of hydroids, hydra, and the Portuguese man-of-war. Hydrozoans can be either polyps or medusae.

**Invertebrate** (in-VERT-uh-brate)—An animal without a backbone.

**Larva** (LAR-vah): **plural; larvae** (LAR-vee)—An animal's early life stage in which its appearance and way of life are very different from that of the adult.

**Medusa** (muh-DOO-suh): **plural; medusae** (muh-DOO-see)—A bell- or umbrella-shaped form of Cnidaria. Adult moon jellies and sea wasps have a medusa form.

**Nematocysts** (NEM-at-uh-sists)—The stinging cells of jelly animals. Nematocysts are living weapons that are used to capture food and for protection.

**Parasite** (PAR-uh-site)—An animal that lives in or on other organisms, taking food from them but giving nothing in return.

**Photosynthesis** (foh-toe-SIN-thuh-sis)—Process by which plants convert sunlight, carbon dioxide, and water into energy in the form of sugars, starches, and other foods.

**Polyp** (POL-up)—A form of Cnidaria made up of a soft stalklike body topped by a mouth surrounded by tentacles. Corals, sea anemones, and juvenile medusa jellies have a polyp form.

**Scyphozoans** (si-fuh-ZO-uns)—A subgroup within Cnidaria made up of moon jellies, sea wasps, and other jelly animals with a medusa form.

**Symbiosis** (sim-bee-OH-sis)—The living together of two different organisms, in which one or both benefit from the relationship.

**Tentacles** (TEN-tuh-kuls)—Long flexible structures around the mouths of medusae, sea anemones, and other jelly animals; used for grasping and stinging.

**Vertebrate** (VER-tuh-brate)—An animal with a backbone.

**Water column**—The area of the ocean between the surface and the bottom.

**Zooplankton** (zoh-uh-PLANK-tun)—Tiny, often microscopic, floating animals eaten by sea jellies.

# Bibliography

**Books and References**

Barnes, Robert D. *Invertebrate Zoology*, 3rd ed. Philadelphia: W.B. Saunders Company, 1974. This authoritative textbook (not light reading) is the bible for learning about invertebrates, for understanding both their classifications and behaviors.

Berril, Michael, and Deborah Berril. *The North Atlantic Coast: A Sierra Club Naturalist's Guide*. San Francisco: Sierra Club Books, 1981. A field guide to coastal and marine habitats and organisms from Cape Cod to Newfoundland.

Grzimek, Bernhard. *Grzimek's Animal Life Encyclopedia*. Vol. 1. New York: Van Nostrand Reinhold Company, 1972.

Halstead, Bruce W., M.D. *Poisonous and Venomous Marine Animals of the World*. Princeton: The Darwin Press, 1978. Sea jelly venoms and their effects. Very technical, authoritative text.

Halstead, Bruce W., M.D., Paul S. Auerback, M.D., and Dorman R. Campbell. *A Color Atlas of Dangerous Marine Animals*. Boca Raton: CRC Press, Inc., 1990. Descriptions of sharks, rays, and jellies; the mechanics of the danger, suggested prevention and treatment, with photographs.

Niesen, Thomas M. *The Marine Biology Coloring Book*. New York: Barnes & Noble Books, 1982. Coloring book–type illustrations of marine organisms with adult text describing anatomy, physiology, and specialized behaviors.

Robbins, Sarah Fraser, and Clarice Yentsch. *The Sea Is All About Us*. Salem, MA: The Peabody Museum of Salem and The Cape Ann Society for Marine Science, 1973. An illustrated field guide to the marine environments and organisms of New England waters.

**Magazine Articles and Periodicals**

"A Most Ingenious Paradox." Stephen Jay Gould. *Natural History*, Dec. 1984. A detailed discussion of the Portuguese man-of-war, especially its colonial character.

"Beware the Jellyfish." George S. Fichter. *International Wildlife*, Sept.–Oct. 1976. Sea jelly physiology and behaviors.

"Clinical Manifestations of Jellyfish Envenomation." Joseph W. Burnett. *Hydrobiologia*, 216/217: 629–635, 1991.

"Jellies with Jaws." Sidney Tamm and Signhild Tamm. *MBL Science*, Vol. 4, No. 1, Spring 1990. A detailed look inside the mouth of the comb jelly, *Beroë*.

"Jellyfish Aren't Out to Get Us." Shannon Brownlee; Sally Drost, Reporter. *Discover*, Aug. 1987. The mechanisms and effects of stinging in sea jellies and their relatives.

"Mechanics of a Turnover: Bell Contractions Propel Jellyfish." Virgil N. Argo. *Natural History*, Aug.–Sept. 1965. A discussion of sea jelly movement, focusing on *Cassiopeia*.

"Sex (and Asex) in the Jellies." Katherine A. C. Madin and Laurence P. Madin. *Oceanus*, Fall 1991. An interesting, technical discussion of salp and other jelly reproductive strategies.

"Transient Jewels: When the Jellies Visit Monterey Bay." David J. Wrobel. *Sea Frontiers*, March–April 1990. Overview of jelly animals found in Monterey Bay.

## Personal Communication

Dr. Laurence Madin, Woods Hole Oceanographic Institution, April 29, 1992.

Dr. George Matsumoto, Monterey Bay Aquarium Research Institute/Monterey Bay Aquarium, April 22, 1992.

# For Further Reading

Doubilet, Anne, and David Doubilet. *Under the Sea from A to Z*. New York: Crown Publishers, Inc., 1991.

Jacobson, Morris K., and David R. Franz. *Wonders of Jellyfish*. New York: Dodd, Mead & Co., 1978.

Tayntor, Elizabeth, Paul Erickson, and Les Kaufman. *Dive to the Coral Reefs*. New York: Crown Publishers, Inc., 1986.

# Index

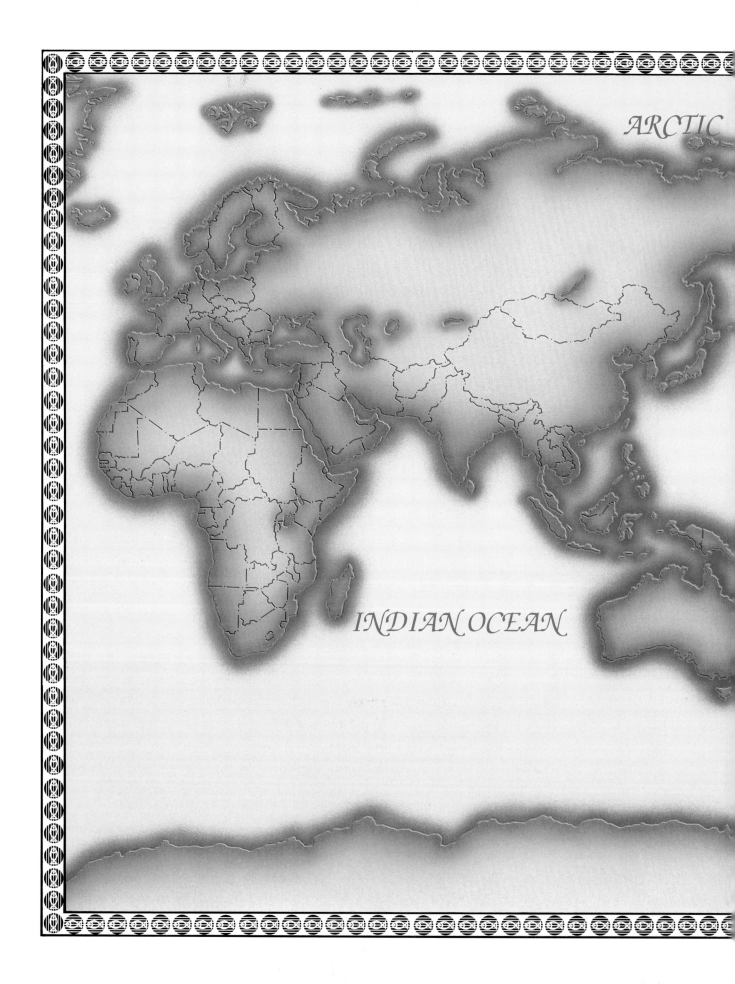

ARCTIC

INDIAN OCEAN

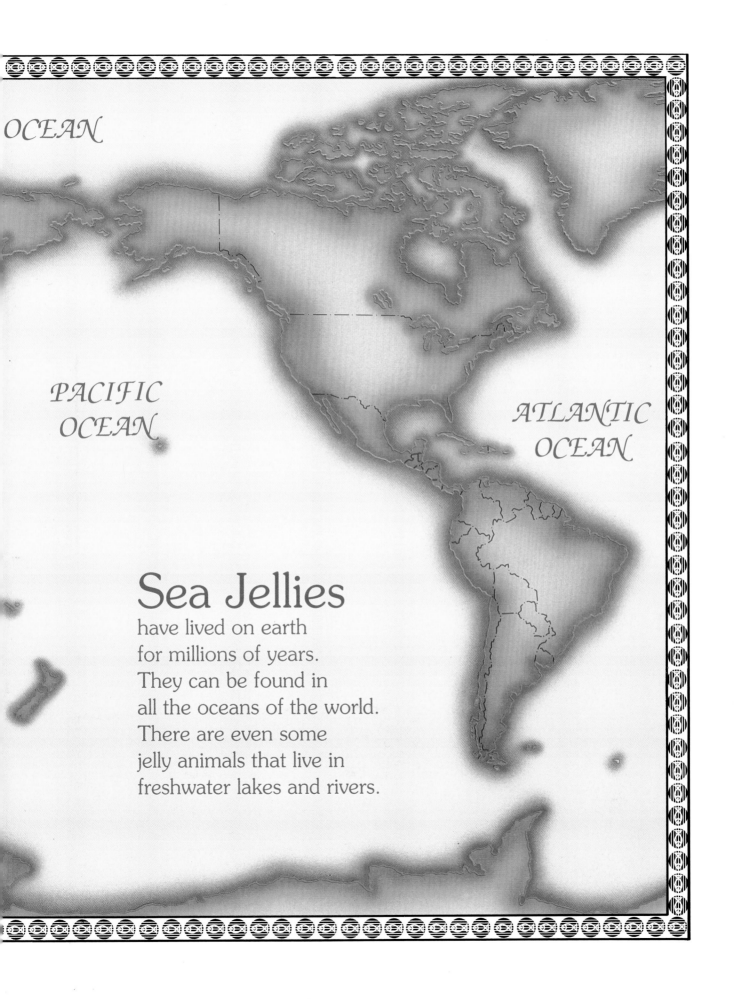

OCEAN

PACIFIC
OCEAN

ATLANTIC
OCEAN

# Sea Jellies

have lived on earth
for millions of years.
They can be found in
all the oceans of the world.
There are even some
jelly animals that live in
freshwater lakes and rivers.